Frequently Asked Questions

all about kava

EARL MINDELL, RPh, PhD

AVERY PUBLISHING GROUP

Garden City Park • New York

The information contained in this book is based upon the research and personal and professional experiences of the author. They are not intended as a substitute for consulting with your physician or other health care provider. Any attempt to diagnose and treat an illness should be done under the direction of a health care professional.

The publisher does not advocate the use of any particular health care protocol, but believes the information in this book should be available to the public. The publisher and author are not responsible for any adverse effects or consequences resulting from the use of any of the suggestions, preparations, or procedures discussed in this book. Should the reader have any questions concerning the appropriateness of any procedure or preparation mentioned, the author and the publisher strongly suggest consulting a professional health care advisor.

ISBN: 0-89529-905-4

Printed in the United States of America

10 9 8 7 6 5

Contents

Introduction

An estimated fifty million American adults—up to 23 percent of women and 17 percent of men—suffer from anxiety disorders, and 9 million Americans have regular bouts of depression. Perhaps you're one of them. Drugs for the treatment of these problems put billions of dollars in the coffers of pharmaceutical companies every year, but these drugs often cause more problems than they solve, and they all have negative side effects.

There is, however, a natural and safe herbal product to improve your mood and reduce anxiety and nervous tension. It's called kava-kava, or simply kava. I'll tell you a lot more about it in the remaining chapters of the book. But first, let's think about why you might want to try it.

Modern day-to-day life is filled with stressors. Ultimately, living with these stressors day after day may cause one to develop an anxiety disorder. Some symptoms of anxiety disorders include unreasonable nagging worries, excessive feelings of fear or foreboding, insomnia, hyperventilation, rapid heartbeat and shortness of breath, fatigue, irritability, and having difficulty concentrating.

Depression isn't so much the flip side of anxiety as it is a bosom buddy. In fact, anxiety is a symptom of depression. When you're afflicted with depression, you can become so overwhelmed with sadness and lack of purpose that you have trouble coping with normal everyday demands. It becomes difficult to have healthy, nurturing relationships and to do satisfying work. Depressed people tend to feel terrible about themselves and may want to withdraw from the world, which only makes a bad situation worse. Often they turn to substance abuse or other self-destructive behaviors to try to escape their pain.

Today, more prescription drugs are being given out for the treatment of anxiety and depression than ever before. Nearly 2 billion dollars worth of the antidepressant medicine Prozac is sold every year. Unfortunately, Prozac and other antidepressant drugs, as well as antianxiety medications, cause several unwanted side effects in the body. There is, however, a safer alternative: a totally natural substance with a centuries-long history of safe use that could ease anxiety and pick you up from depression. Wouldn't you want try it before resorting to the use of expensive, risky prescription drugs?

If your answer is "yes," you've picked up the right book. In these pages, you'll find everything you need to know about kava, a remarkable herb that soothes anxiety and lifts depression. Kava is a natural remedy for anxiety, including symptoms of nervousness, panic, heart palpitations, stomach irritation, irritability, and fatigue. It can help in depression and insomnia, and many other physical and behavioral problems.

Most of the modern research on kava has come from German laboratories. As a result of this research, kava-kava has become a very popular remedy for anxiety and depression in Germany, Austria, England, and Switzerland. Now, the word is spreading around the world.

All About Kava will give you all the information you need to use kava safely and effectively. You'll learn about the fascinating history of kava and find out exactly what to expect when you take it. You'll understand how to use kava for anxiety—by far its most popular use. And you'll learn about possible interactions with other drugs you might be taking, so you can use kava safely.

Life without the burden of anxiety and depression is possible, and kava can be an important step towards lifting that burden from your shoulders.

1.

A Brief History of Kava

Westerners first took note of kava during Captain James Cook's 1768–1771 voyage to the South Pacific. South Pacific Islanders are known for their tolerant, easygoing ways. Although life on a lush tropical island thousands of miles from civilization pretty much rules out the existence of anxiety and depression in these cultures, the regular use of kava root beverages brings their friendliness and contentment to a higher level. Native islanders discovered centuries ago that the root of the kava plant created a sense of blissful well-being in those who ate it or drank a ceremonial drink containing it.

It has been a part of religious, community, and social occasions on these islands ever since, much like alcoholic drinks have become part of social rituals in America and Europe.

Q. What is kava?

A. The kava plant, known to scientists as *Piper methysticum*, grows throughout the beautiful islands that dot the South Pacific, including Hawaii, Papua New Guinea, Fiji, Samoa, and Tahiti. Like garlic, it no longer grows wild and must be cultivated. Kava has been an important part of the people and their many cultures on these islands for the past three thousand years.

The word "kava" means bitter, sour, pungent, or sharp. The plant is also known as *kava-kava, awa, ava* in Hawaii and Tahiti, *gea* in the Banks Islands, *gi* in the Torres

Islands, and *yaquona* in Fiji. Its Latin, or scientific, name is *Piper methysticum*, meaning "intoxicating pepper." The kava plant is a leafy shrub with many branches that reaches maturity about three to five years after planting. It can grow to a height of eight to ten feet, and it grows best at altitudes slightly higher than sea level. Kava's leaves are broad and heart-shaped, and its flowers are long and thin, barely noticeable and without visible petals. The roots grow into thick cords or a large, tangled mass. There are several varieties of the plant, some of which grow wild. The one described in these pages is a cultivated version that has to be planted. It's found in Papua New Guinea, the Micronesian Islands (Ponape, Kosrae, Soloman), the Melanesian Islands (Wallis, Fiji, Samoa, Tonga, and a chain of twenty-two small islands collectively known as Vanuatu), and the Polynesian Islands (Tahiti and Hawaii).

Q. What are some of the legends about kava?

A. If you ask a resident of a region called Manu'a about the origins of the kava plant, he might tell you that it all began with a god named Tangaloa who decided to visit the earth on a fishing trip and wanted some kava to drink with the fish he caught. According to this legend, Tangaloa sent two mortals to heaven to bring him a root, but they mistakenly brought the whole plant. As the god pulled the roots away, he scattered the other parts all across the land and new kava plants sprouted wherever those parts fell.

If you ask a resident of Tonga the same question, she might tell you that a great chief, expected to serve a feast to a god in a time of famine, killed his only daughter and cooked her. In this version, the god immediately recognized that he was being served human flesh, and told the chief to bury his daughter's remains and to bring him the plant that grew from that spot. When the chief did so, the god showed him how to prepare a ceremonial drink from it.

Q. How were these ceremonial drinks prepared?

A. In the late 1700s, as a member of the team of explorers led by Captain James Cook, European naturalist George Forster first reported on how natives of these tropical islands used the roots of this tall, leafy plant. These explorers reported that rituals and social gatherings centered around the preparation and consumption of a kava drink made with ground roots and water, served in a halved coconut shell. Sometimes, the roots were chewed by specially designated members of a tribe, usually young girls, who took great care to prevent wetting the root with their saliva. Those designated with the task of chewing kava didn't get to partake of the drink once it was prepared. Only those who were young and free of disease could perform this important task, and it's said that those who had to chew kava for their elders complained of how difficult and tiring it was. Folklore tells us that many young girls eloped to escape kava-chewing duty.

When the Europeans arrived on the scene, they found this chewing technique to be repulsive. After colonizing the islands, they outlawed it wherever they could. Chewing of the roots has been all but replaced by grinding with mortar and pestle, and the kava you buy in the health-food store has probably been prepared with machines that grind it into a fine powder. Researchers have found, though, that mixing kava root with saliva increases its potency—so the islanders were on the right track when they chose to prepare it this way!

Q. What were these kava ceremonies like?

A. Many different occasions, including the greeting of respected visitors, town meetings, and informal gatherings, are still marked by the ceremonial drinking of kava. Former President Lyndon Johnson and his wife Lady Bird Johnson quaffed cups of it on visits to the South Pacific. Pope John Paul II sipped kava in a ceremony welcoming him to an official Fijian gathering. Queen Elizabeth also enjoys a cup of kava when she visits Fiji. Hillary Rodham Clinton was given kava by a Samoan community on Oahu, Hawaii, during her husband's 1992 presidential campaign tour. Kava roots are given as peace offerings and ceremonial gifts.

The kava ceremonies of Pacific Islander societies are often very formal. Members of the tribe with the lowest social station prepare the kava drink, and the order in which they serve the drink to the others reflects the social or political station of each. The most prestigious people—chiefs and visiting dignitaries—are served first, and so on down the hierarchy. Each cup of this brew contains far more kava than I will recommend at a sitting, and such a generous portion brings on a powerful state of relaxation.

European explorers reported seeing these islanders enjoying a pleasant, contented, tranquil feeling after drinking kava, sometimes drifting off to sleep and waking up refreshed, without a hangover. Of course, puritanical missionaries saw this as unacceptable revelry and tried to outlaw kava, but in most cases the traditions held.

Q. What are the active ingredients of kava?

A. Anywhere from 3 to 20 percent of the kava root is made up of kavalactones (also known as kavapyrones), the active ingredients of the kava plant. These compounds include demethoxy-yangonin, dihydrokavain, yangonin, kavain, dihydromethysticin, and methysticin.

For the past 130 years, scientists have carefully studied each of these chemicals to determine how kava works in the body. Methysticin and dihydromethysticin seem to protect the brain from injury due to stroke. Kavain, dihydrokavain, methysticin, and dihydromethysticin are the painkilling ingredients of kava. Dihydrokavain and dihydromethysticin appear to be responsible for the sleep-inducing, soothing effects of the root. Kavain, dihydromethysticin, and methysticin relax smooth muscle—the kind found in the walls of the intestines and arteries—and prevent convulsions.

As with most botanical medicines, the

purified active ingredients used alone don't work as well as the extract containing the whole plant. Most studies have used a standardized extract containing 70 percent kavalactones. This means that for every 100 mg of kava, 70 mg is composed of kavalactones, and the rest contains other important substances that enhance the body's use of these active ingredients. Brain absorption of kavalactones as part of kava extract is up to twenty times better than that of the isolated active ingredients. Furthermore, when whole root extracts are used rather than isolated kavalactones, there's an additional benefit: each of the kavalactones appear to interact with the others in a synergistic way.

The roots of kava plants vary a great deal in the amount of the active ingredients they contain. You should use a standardized extract, which ensures that it contains enough kavalactones to have the effects you're taking it for. You can buy kava standardized for 30 to 70 percent kavalactones.

According to Michael Murray, N.D., a leading authority on the subject of natural medicine, the use of less concentrated extracts may actually be best. A 70-percent extract may contain too few of the other important ingredients contained in the root for your body to make good use of the kavalactones, while a 30-percent extract has a composition more similar to that of the plant itself.

Q. What are kavalactone extracts?

A. Kavalactones are the key active ingredients in the kava plant. When you buy kava supplements, the label should identify the percentage of kavalactones in capsules.

Q. How does kava work?

A. Most theories about depression and anxiety say that they're caused by an imbalance of brain chemicals called neurotransmitters. Most antidepressant and anti-anxiety treatments—herbal and pharmaceutical—work by affecting the actions of these chemicals.

Researchers don't yet understand the mechanism by which kava relieves anxiety. There are a couple of theories, one or both of which could explain kava's effects. One idea is that they clear the area around the receptor sites for a brain compound called gamma-amino butyric acid (GABA), so that more of the neurotransmitters can connect to the receptors. The more GABA binds to receptors, the more relaxed we become.

Another idea is that kava somehow acts on a part of the brain called the amygdala. The amygdala are two chickpea-sized lobes located in a part of the brain called the limbic system. Amygdala are the parts of the brain that are thought to be functioning improperly when anxiety hits. Neurotransmitters are out of balance in these tiny lobes during bouts of anxiety.

The limbic system is the part of the brain that controls emotions, instinct, and basic body functions like heart rate, blood pressure, body temperature, blood sugar levels, sex drive, sleeping, waking, and appetite. This would help to explain kava's far-reaching effects, as the limbic system is the "control center" for all of these body functions.

Q. Will kava make me sleepy or unable to think clearly?

A. Native drinkers of kava can perform just as well on tests of reaction time and thinking ability when they've drunk kava as when they haven't. With high doses, there is some alteration in vision, with the pupils dilating and some aspects of focusing becoming difficult.

In one study, a group of students was given six pints of kava beverage over a two-hour period. This is an enormous dose, which I do *not* recommend. The students seemed sleepy, slurred their words a bit, and had bloodshot eyes—but they could still walk a straight line and dash up a set of stairs two at a time.

Kava doesn't intoxicate like alcohol. Amazingly, it even improves thinking abilities and doesn't cause sleepiness if taken in the right amounts. In two studies comparing kava with the prescription anxiety drug oxazepam (similar to the tranquilizer Valium), kava was equally effective in relieving anxiety and slightly improved scores on mental function tests. Oxazepam decreased attention span and ability to process information; kava improved performance on the same tests.

Q. What is the difference between using an herbal remedy like kava and using a prescription drug?

A. Prescription drug manufacturers originally find many of their products in plants and other natural substances. Then they try to isolate the "active ingredients," or alter the natural substance so that it is no longer identical to anything that occurs in nature.

This is done so that the drug can be patented. Patented drugs are worth far more money than any natural substance. Drug companies holding patents for new drugs stand to make a fortune, because they can charge whatever they like. It's much harder to obtain a patent on a natural substance, such as kava, and so there is free market competition between those who wish to sell it. Prices are kept low.

In addition, prescription drugs tend to

be very powerful in their action, and as a result they all have undesirable side effects. Herbal medicines tend to be more gentle and rarely have side effects when used as directed by an experienced health-care professional.

Researchers who try to isolate the "active ingredients" of plant medicines to better understand how they work discover time and again that the isolated ingredient doesn't work nearly as well as the whole plant. The various parts of the plant interact with one another to do their healing work, gently bringing our bodies back into balance. Mother Nature's medicines have a synergy all their own that is rarely, if ever, found in prescription drugs.

2.

Kava for Reducing Anxiety

Feeling anxious or tense? Worried about losing your job? Braced for another argument with your spouse? With the combined pressures of work and home, and less quality time for the things you really want to do, anxiety has become a common illness nationwide. While a glass of wine, a beer, or hard liquor can help you relax, most experts worry about the long-term effects of alcohol on health. The same applies for prescription drugs, such as tranquilizers. Kava is a natural remedy for anxiety and has passed the test of time.

Q. What is anxiety?

A. Anxiety is a generalized feeling of worry, nervousness, or apprehension about the future. Although anxiety is sometimes a normal response to people and situations, it should be only temporary. Feeling anxious all the time is a serious stress that can affect your overall mental and physical health. It can influence how you think about people and situations, and, perhaps, wrongly so. Being anxious also affects your entire body—your skin, your blood vessels, your heart, your muscles, and your digestion.

Some researchers think that chronic anxiety and depression are the result of an imbalance of certain neurotransmitters, which are substances that communicate information in the brain. When levels of these neurotransmitters are high, the body reads them as signals to tense up in preparation for fight or flight. When they are low, depression strikes.

Q. What is the relationship between anxiety and stress?

A. Anxiety is the result of your body's response to stress, and it doesn't matter if the stress is real or imaginary. Although we often think of stress as a bad thing, our bodies' responses to stress are actually healthy under many circumstances. For example, if you are crossing the railroad tracks and suddenly you see a train barreling towards you, your senses send the message to your brain: "Get out of the way!" The brain sends messages to the glands that make certain hormones, and these hormones travel quickly throughout the body, sending messages to the organs and the muscles to prepare for quick action. The heart beats faster and blood pressure rises, sending extra blood flow to the muscles and away from the organs that aren't part of the escape effort. All of these changes work in concert to create a burst of energy so you can get out of harm's way.

In this day and age, the stress response can be counterproductive. Many of the stresses we encounter are not the kind we can respond to by dashing out of the way. Our lives are filled with little dilemmas and worries that can wind us up as tightly as can major stressors, and our bodies really don't know that these stresses are different from those that threaten our physical health. Blood pressure goes up and stays up, the heart beats rapidly, the walls of the intestines and colon cramp, and muscles stay tense.

Q. How will kava help anxiety?

A. When South Pacific Islanders drink the

traditional mix of kava root and water, they become more sociable and relaxed. Kava drinking has been compared with wine drinking in European cultures—both tend to bring out feelings of contentment, well-being, and release from stress—but kava doesn't cause the same kind of drunkenness that alcohol causes. Some users report feelings of being more "in the moment" and able to focus on the present rather than worrying about the past or the future. Senses are sharper, and sensitivity to noise and light increases.

The amount of active ingredient in kava drink is much greater than what's needed for relief of anxiety or depression, and small doses may not have very pronounced effects. You might notice only relief from anxious feelings or deep sadness as kava brings you back into balance. Situations or people that once caused you to fly into a panic or sink into depression might not affect you as strongly when you use kava.

Q. How long does it take for kava to relieve anxiety?

A. When taken in large doses, kava has almost immediate effects. If you use it in smaller amounts, you may need to give it a few days to have an effect. In one study on fifty-eight patients with anxiety disorders, a low dose (100 mg of kava, containing 70-percent kavalactone extract three times daily) significantly improved anxiety symptoms in only one week's time—with absolutely no adverse reactions. Most other studies have shown improvement in anxiety within four to eight weeks.

Q. How much kava should I take for anxiety?

A. You will likely have to experiment a little to find the minimum amount that works.

In general, you will probably want take from 70 to 210 mg of kavalactones daily, divided into three equal dosages. You can take one capsule with each meal. Depending on the strength of the extract you're using, you may need to do a little math to figure out how many capsules to use. Many kava supplements contain 100 mg of extract containing 70 mg of kavalactones; if the one you choose fits this description, simply take a capsule three times a day.

If you use a 100 mg supplement with 30 percent kavalactone extract, you may want to take two at each meal (for 60 mg kavalactones per dose). Don't exceed the recommended dose, because doing this will probably make you feel sleepy. If three 100 mg capsules containing 30-percent kavalactone daily works, stick with that. If it doesn't, gradually increase the dosage until you get relief.

Q. Should I take kava every day, or should I only use it when I'm feeling very anxious?

A. Anxiety can prevent us from dealing with our problems, and relieving the symptoms can free us up to get ourselves back on track. Be careful, though, not to let kava become an excuse for not correcting those problems. For example, it's probably better to change a bad job than to take kava. While kava is harmless when used for long periods, you should not use it as an escape from everyday life.

Use kava for a week to a month at a time when your anxiety is interfering with your ability to function day to day. Try to use that time to work on fixing whatever has been making your anxiety unbearable. Then you can stop using kava for a while and see how you do.

Q. Will kava really make me more sociable?

A. Yes—it's used for this reason in the South Pacific, much like alcohol is used in Western countries. If you have considerable anxiety about being in group situations, kava can help you overcome your fears and enjoy yourself more. A reader wrote to me that he sometimes felt paralyzed by shyness in social situations and usually hated going to parties. His work required him to go to business dinners and parties fairly often, though, so he tried taking 100 mg of 70-percent kavalactones kava extract about one hour before going. He found that this made it much easier to interact comfortably with others. "I don't put my foot in my mouth nearly as much," he wrote," and I'm gaining the confidence I need to overcome my shyness. If I can be witty and charming with the kava, I certainly can be witty and charming without it!"

Q. Is kava a better anxiety remedy than drugs like Valium or Xanax?

A. The class of drugs known as benzodiazepines (including the popular antianxiety drug Valium) mimic the actions of a neurotransmitter called gamma-aminobutyric acid, or GABA. This has a soothing effect on the body, and induces relaxation. These drugs are part of most prescriptions for the treatment of anxiety. Xanax, Valium, and Ativan are commonly prescribed benzodiazepines.

These drugs tend to cloud thinking, slow reaction time, and cause sleepiness. They are addictive, and those who take them for long periods develop a tolerance to them. Larger and larger doses become necessary. When the drugs are stopped, there are usually serious withdrawal symptoms to contend with. Anxiety, trembling, low blood pressure, agitation, insomnia, dizziness, loss of appetite, ringing in the ears, blurred vision, diarrhea, and even psychosis and seizures are all possible symptoms of benzodiazepine withdrawal.

Unlike these drugs, kava relieves anxiety and depression without side effects, gently balancing body chemistry rather than manipulating the action of brain chemicals.

Q. What are some other ways I can relieve anxiety?

A. Kava isn't a cure for the problems of chronic anxiety. It isn't meant to be used indefinitely and shouldn't be regarded as a way to avoid fixing the problems that are causing the feelings. Until deep healing becomes possible, you can use this herbal remedy to improve your function and your ability to cope with day-to-day stress. If you are anxious or depressed in response to stressful situations or difficult events in your life, you can use kava to get through these times more easily.

Isn't it amazing how some people can stride through the whirlwinds of work, family, and everything else, and still stay cool as cucumbers? Those people aren't necessarily under any less stress than those who crumble at the slightest hint of adversity; they simply have found effective methods for coping with it.

In *The Relaxation Response*, Herbert Benson, M.D., describes methods anyone can use to relieve the physical symptoms of anxiety. Simply breathing deeply while seated with eyes closed for ten to twenty minutes brings heart rate and blood pressure down, relaxes tense muscles, and relieves feelings of anxiety very effectively.

Using relaxation techniques or doing any kind of meditation can give you power over your body's responses to stress. Yogis in deep meditation have the ability to lower their body temperatures and heart rates to the level of a hibernating animal. This kind of control allows the yogi to elicit a relaxation response in the body whenever he decides to do so.

Meditation and yoga are excellent practices for learning to cope with anxiety. For the average person, this translates to being able to manage our stressful lives, even as we encounter the usual obstacles—and having control over our bodies' responses to those obstacles.

Q. How can I apply some of these techniques, and are they compatible with using kava?

A. When you're driving in your car and someone cuts you off, almost causing an accident, your heart beats faster, you break into a sweat, every hair on your head and your body feel like they are standing up, and you begin to shake. Many people respond to this by completely losing their cool and shouting unrepeatable epithets at the other driver. Your reaction may put you at risk for having an accident, but it also damages your body. This much anger puts unnecessary wear and tear on the heart and blood vessels. The situation causes huge surges of stress hormones that, over long periods of time, can cause damage to the brain and muscles.

If you have been practicing relaxation techniques or yoga—a combination of stretching, concentration, strengthening, and meditation—you will find that by simply taking some deep breaths and easing into a relaxed state, you will avoid these compounding stress responses. After a time this response to anxiety becomes automatic. And yes, taking kava is compatible with these relaxation and de-stressing techniques.

Q. What else can I do to deal with severe or chronic anxiety?

A. If you are debilitated by chronic anxiety or depression that seems unrelated to events in your life, you may need to get some psychological counseling. Ask around for recommendations for a good therapist and invest a little time and money in your own well-being.

You may also need to make life changes, such as quitting a job you hate or severing a relationship that isn't working. Even if you feel trapped, there's probably a way out you haven't seen yet. It isn't ever worth it to stay where you are if it's making you ill.

As an example, a friend of mine complained to me that she woke up every single night in a panic, and that her days were interrupted by attacks of paralyzing fear. She was in an abusive marriage and didn't want to leave her husband. She tried Prozac, Valium, St. John's wort, kava, and finally counseling. Fortunately, the counselor saw that my friend was having these attacks of anxiety because she was in a bad marriage, and he helped her to recognize this. As soon as she left the marriage, her once debilitating anxiety attacks went away.

Eating a well-balanced, plant-based diet and drinking plenty of clean water provide the groundwork for a less anxious existence. Avoid junk food, fast foods, and other foods that are highly processed. If you can, get some exercise every day—take a twenty-minute walk after work or start training for a 100-mile bike race. Whatever feels best to you is the best kind of exercise.

And, importantly, give yourself time to do what you want to do. Without that, you'll never feel free of the crushing stresses of life. No matter what, find a few minutes to a few hours as often as possible to do what you love to do. Watch movies, go skiing, sit in a café, and scribble in your journal. Get a babysitter if you have to, or call in sick to work when you need some time. Nothing is more important than your health, and if you're operating on overload all the time, you're going to be pretty ineffectual anyhow. And, of course, use kava when it all gets the best of you.

All of this is good, sound advice whether you take kava or not.

3.

Kava for Depression and Insomnia

Are you feeling down a lot of the time and see no hope of feeling better? Are you having difficulty sleeping? These are also signs of mood disorders—particularly depression and insomnia. Kava can elevate your mood, and it can also help you get more restful sleep.

Q. How do I know if I'm depressed?

A. Feeling sad on occasion is a normal part of our emotional lives, and we shouldn't try to avoid it completely. For example, it's reasonable to feel sad when a loved one, even a favorite pet, has died. Many people will sometimes say they feel depressed when they're feeling blue, if things aren't going their way, or if they're temporarily experiencing sadness. This is very different from the clinical diagnosis of depression, which is a combination of feeling very sad and of having no expectation that things will ever get better.

Some people who suffer from depression describe it as sadness, but others describe it as numbness or inability to feel anything at all. Mild, long-lasting depression (*dysthymia*) can last for two or more years, while major depressive episodes are usually short-lived but devastating. Any variety of depression makes it very hard to enjoy life.

Symptoms of depression can include:

- Persistent sadness and/or anxiety.
- Feelings of emptiness, worthlessness, and guilt.
- Hopeless and pessimistic feelings.
- Irritability, restlessness, and difficulty in thinking and making decisions.
- Headaches, digestive problems, or pain that can't be traced to any physical malady.
- Loss of libido (desire for sex).
- Insomnia or sleeping too much (hypersomnia).
- Loss of appetite, or dramatic increase in appetite, with weight loss or weight gain.
- Fatigue.
- Suicidal thoughts.

Q. How does kava help with depression?

A. Because two out of every three people who are depressed also suffer from anxiety, the two conditions are clearly intertwined. Symptoms of depression—including irritability, anxiety, chronic pain, and loss of libido—can be remedied by kava.

Q. How is kava different from St. John's wort in its effectiveness in treating depression?

A. St. John's wort (*Hypericum perforatum*) is a flowering plant that has effects similar to those of the prescription drugs known as serotonin reuptake inhibitors (SSRIs), such as Prozac, Zoloft, and Effexor: They block the reabsorption of serotonin, keeping it active longer. Serotonin is a neurotransmitter, or brain chemical, that promotes relaxation. St. John's wort also has similarities to antidepressant drugs called monoamine oxidase inhibitors (MAO inhibitors), which keep levels of mood-lifting neurotransmitters elevated.

Like kava, St. John's wort works as well as prescription drugs, and it doesn't adversely affect thinking ability or pose any risk of addiction. To benefit from St. John's wort's antidepressant effects, however, you must take the recommended dose every day for at least one month. This isn't the case with kava, which has almost immediate effects. Also, St. John's wort can sometimes make anxiety worse while it relieves depression.

St. John's wort also has potentially harmful interactions with certain foods. Smoked or pickled foods are high in the amino acid tyramine, which can cause dangerous blood pressure increases when mixed with St. John's wort. The amino acids tryptophan and tyrosine, both of which can be taken as supplements, have the same effect. Alcoholic beverages, cold or hay fever drugs, amphetamines ("uppers"), and narcotics ("downers") can also have dangerous interactions with St. John's wort, causing blood pressure to shoot up. Kava doesn't have any known interactions with foods, although using it with alcohol or other recreational drugs can certainly make you unable to think straight or operate a car.

If your main problem is depression, you can try St. John's wort. If your main complaint is bouts of anxiety, kava is your best bet.

Q. How does kava compare with antidepressant drugs like Prozac?

A. Prescription drugs like Xanax and Prozac—prescribed for anxiety and depression, respectively—are artificial versions of substances found in nature, altered to try to make them very specific in their actions in the body. Side effects are common with these drugs because you can't target the workings of a single system without affecting others.

The antidepressant Prozac is a serotonin reuptake inhibitor (SSRI). Other drugs in this class include Paxil, Zoloft, and Effexor. The downside of SSRIs is their side effects. Prozac, for example, can cause itchy skin rashes, headaches, insomnia, nervousness, and lack of sex drive. In some cases, high levels of serotonin can cause "serotonin syndrome," with chills, fever, muscle spasms, agitation, confusion, and feelings some describe as "electrical currents" running through them.

Prozac even causes some people to become antisocial, intensely suicidal, or violent towards others. There have been at least seventy court cases in the United States so far where defendants have blamed uncharacteristically violent, murderous, or suicidal behaviors on Prozac—the "Prozac defense."

Complaints to the Food and Drug Administration (FDA) about Prozac number almost 40,000—the highest number of complaints about any drug in FDA history. Four SSRIs (Prozac, Paxil, Zoloft, and Effexor) rank in the FDA's top twenty list for number of reported adverse effects. These drugs aren't anywhere near as risk-free as we've been encouraged to believe.

The majority of people who suffer from depression also suffer from anxiety, and yet SSRIs can cause anxiety. Kava safely relieves both depression and anxiety.

Q. Can I stop taking SSRI drugs when I start taking kava?

A. If you are using an SSRI, do not stop taking it abruptly. Withdrawal symptoms, including nausea, vomiting, pain, insomnia, uncontrollable crying, anxiety, chronic fatigue, nightmares, and suicidal thinking, have afflicted those who have stopped these drugs suddenly. SSRIs change brain chemistry dramatically, and abruptly withdrawing them throws the balance of neurotransmitters way off. Taper the dose down as gradually as possible, under the supervision of a health-care professional.

I'll give you an example. A fifty-five year old reader wrote to thank me for helping her get through the unexpected death of her husband. For weeks after his death, she had felt completely paralyzed, unable to sleep, work, eat, or even handle the arrangements for his burial. She had constant headaches and stomach pains and couldn't stop crying. As months began to pass, she realized she could no longer live this way, and she went to her doctor, who gave her a prescription for Prozac. It had helped her for a while, boosting her mood enough to get her functioning, but soon she began to feel anxious and jittery. She had no appetite and was seriously underweight. Suddenly she felt she didn't want to be around anyone, not even her own children.

"I read something you wrote about the dangers of Prozac," she told me, "and I recognized so many of those side effects, I decided to quit taking it and try kava instead." She tapered down her Prozac dose, and soon after taking her last dose, she began to use a combination of St. John's wort for depression and kava for sleeplessness. She feels that she's back on track now and successfully putting her life back together.

Q. What about insomnia? Is kava better than sleeping pills?

A. You might think that depression would encourage a person to sleep. However, anxiety and depression can disturb sleep patterns. Poor sleeping patterns lead to a downward spiral of constant tiredness, and insomniacs lay awake much of the night and spend their days in an exhausted fog.

In many cases, high doses of tranquilizers—usually benzodiazepines like Valium or Xanax—are prescribed as sleep aids. Sleeping pills are addictive, and your body builds up a tolerance so that higher and higher doses are needed to have the effect of sending you into a restful night's sleep. People who use tranquilizers and sleeping pills on a regular basis usually end up dependent on them—they can't get to sleep without them, and if they stop taking them their insomnia becomes worse than ever. Kava can bring about deep, refreshing sleep without the hangover of sleeping pills. That makes it far superior to these prescription drugs as a treatment for insomnia.

Q. How much kava should I take to help me sleep?

A. Try 70 to 210 mg of kavalactones—one to three capsules of 70-percent standardized extract, or two to seven of 30-percent extract—one hour before going to bed. Try to find the lowest effective dose, so start with 70 mg of kavalactones.

Q. What about melatonin? Isn't that supposed to be the best natural cure for insomnia?

A. Melatonin is a hormone secreted by the pineal gland. This tiny gland releases melatonin when darkness falls. "Time to sleep," says this hormonal messenger, and the body rests. Now that our neighborhoods, homes, and workplaces are lit brightly even after nightfall, our bodies often don't know when to make melatonin, and sleep patterns can become disrupted.

Jet lag is another good example of sleep-pattern disruption. When the length of your day changes after you've flown across several time zones, your body's smooth rhythm of melatonin secretion is baffled by night falling too early or too late. Taking some at bedtime in your new time zone jump-starts your sleep patterns to get you back on track. In fact, researchers are excited about melatonin's many other health-promoting roles in the body—some believe it's the key that will unlock the door to the fountain of youth.

As is the case with anything that sounds too good to be true, there are problems with everyday use of melatonin. It is a hormone, and we are only beginning to understand how it works. Researchers are still not certain that using large doses of melatonin for months is safe. The amounts secreted naturally are many times less than the 1 to 3 mg normally used as a sleep aid. On the other hand, the safety of kava has been proven over centuries of use, and won't make you feel hung over.

Q. Is there anything else I can do to help cure my insomnia?

A. If you are having trouble falling asleep, there are changes you can make to naturally enhance your body's production of melatonin. When night falls, keep artificial lighting to a minimum. Use only as much light as you need to be able to read or relax. Avoid fluorescent lighting at night whenever possible. Adjust your schedule so that you can go to bed around the same time every night, and spend some time unwinding at the end of the day. Use your bed only for sleep and spending time with your significant other—don't read, work, or watch television in bed. By getting ready for bed, you help your pineal gland make the connection that when it's time for bed, it's time for sleep.

Also, don't drink coffee or tea or eat chocolate (sources of caffeine) after three in the afternoon. If you like to snack in the evening, try something high in complex carbohydrates, such as whole-grain toast with honey and a dab of butter, or a bowl of whole-grain cereal. Carbohydrates raise levels of an amino acid called *tryptophan*, which has a sedating effect.

Another way to relax your body is to take calcium and magnesium supplements. These are essential dietary minerals that relax muscles and encourage restful sleep. You can buy calcium-magnesium supplements at your health food store and take one an hour before bed.

4.

Kava for Pain and Other Conditions

Analgesic, or pain-relieving, drugs are the best-selling over-the-counter drugs in the United States. People suffer all sorts of pain, from headaches to achy joints. Most headaches are tension-related and, likely, anxiety-related. By reducing anxiety, kava may be able to ease some types of pain. It may also provide some benefit to people with cardiovascular diseases and menopausal symptoms.

Q. Does kava work as a pain reliever?

A. Kava is a powerful local anesthetic. It can also be used to improve the effectiveness of other local anesthetics. No one has figured out how exactly kava works to numb the areas it touches. It doesn't work as a pain reliever like aspirin—it won't necessarily relieve a headache if you take it by mouth (unless your headache was caused by anxiety). It does, however, numb any surface it's applied to. (Think of yourself leaving the dentist's office with a face full of Novocaine if you need clarification on this point.) Hopefully, kava will soon be used as a safe and effective alternative to local anesthetics, or will be added to those now used to improve their effectiveness.

Q. Does kava relax tense or cramped muscles?

A. The muscle relaxant effects of kava have been clearly shown in animal experiments. One of the legends of man's discovery of kava describes a tribesman seeing a rat chew on a root, fall down limply as if dead, sleep soundly for hours, and spring back to life as if nothing had happened. The human observers likely said, "I've got to try this stuff," and a tradition was born.

In many cases, headache pain, some types of back pain, and muscle aches can be traced back to anxiety. Kava can help relieve these types of aches and pains. Anxiety causes muscles throughout the body to become tense. Overstimulation by "fight-or-flight" neurotransmitters and stress hormones tells muscles that they should be ready for action at any moment. This unnecessary muscle tension leads to soreness, muscle cramps, headaches, irritable bowel syndrome (as the muscular walls of the small intestines and colon tighten and cramp), jaw clenching, and even irregular heart rhythms (your heart is just a big ball of muscle, after all). With illnesses that lead to chronic pain, such as arthritis and cancer, the stress of being in pain all the time can cause a snowballing effect, increasing muscle tension and making pain worse. Kava can interrupt this cycle.

Kava relaxes the tightness of overtense and cramped muscles very effectively. Nerve signals from the spinal cord are partly blocked by the active ingredients of the kava root. The effect relaxes the muscles that have been contracted, or tightened.

Q. Will kava work as well as the over-the-counter pain medications I usually use?

A. Analgesic drugs, such as aspirin, Tylenol (acetaminophen), and Advil (ibuprofen), are so commonly used to relieve everyday aches and pains that we tend to think they're harmless. I know people who pop these drugs every day without a second thought.

What I usually tell these people is that these drugs are far from harmless. Most already know that aspirin is hard on the stomach, and that's why they switch over to acetaminophen or ibuprofen. But ibuprofen injures the lining of the small intestine and stomach and can cause dangerous gastrointestinal bleeding, and acetaminophen can be very damaging to the liver. Aspirin and the ibuprofen family of nonsteroidal anti-inflammatory drugs (NSAIDs) are also a major cause of what's called "leaky gut syndrome," where tiny holes are created in the intestinal lining, allowing undigested bits of food into the bloodstream and causing allergic reactions.

If you're taking NSAIDs for arthritis, you should know that these drugs actually inhibit the formation of collagen, the building block of all connective tissue. Without an adequate cushion of cartilage between the bones, they rub together, causing tissue swelling and pain. NSAIDs may temporarily relieve pain and swelling, but in the long run, they make arthritis worse by preventing the healing of damaged cartilage.

The bottom line is that these drugs are okay once in a while, but frequent use isn't necessary or worth the risk. In contrast, kava relieves tension and anxiety, which should help relieve pain.

Q. What else can I do for pain relief?

A. Listen to your pain; it's often a message that you need to take it easy. Taking a few minutes to relax quietly or meditate can often cure headaches just as well as can popping pills. If you have chronic headaches or muscle pains, look into stress-reduction programs that take you through guided relaxation exercises or meditation.

If you're an avid exerciser with a strained muscle or sprained joint, don't take painkillers so that you can exercise sooner—give overtaxed muscles time to heal. Without proper rest, you increase your chances of reinjury greatly.

In addition, your body makes its own painkillers—*endorphins*—when you work out. Take a walk, head to the gym, take a bike ride, or try water exercise. Whatever gets you moving and stretching should be an enormous help, especially if you have chronic back pain or arthritis.

For any kind of inflammation, ice packs work wonders. They decrease swelling and numb the area. Make sure you leave the cold pack on for at least twenty minutes. If you're having muscle spasms, use heat alternated with ice for twenty minutes each.

Q. Can kava help keep me free of cancer and heart disease?

A. Kava is not a treatment for cancer or heart disease. However, it may help correct some of the stress-related habits that increase the risk of these diseases. Scientists are learning more and more about how important a role stress plays in the development of illnesses like heart disease and cancer.

People who suffer from constant anxiety

or depression are at much greater risk of getting sick. Using kava to soothe the body's stress response can help prevent this turn of events.

Q. Can kava help recovery from stroke?

A. Most strokes are caused by a blood clot interrupting the flow of blood, oxygen, and nutrients to part of the brain. Stroke can cause death, and it can be devastating for those who survive, as parts of the brain needed for clear thinking, memory, and movement are destroyed. Ministrokes are the second leading cause of senility.

In one experiment, researchers induced strokes in laboratory animals and found kava to be neuroprotective—meaning that it keeps those parts of the brain alive during a stroke that might otherwise die for lack of oxygen. When these rats were given kava extract or isolated kavalactones before a stroke, much smaller portions of their brains were destroyed than when they weren't given the kava first.

Q. Is kava a good supplement for menopausal women?

A. Anxiety is a hallmark of menopause. A woman's body goes through enormous changes during this time, and the emotional and physical discomfort can be hard to handle. In one study, forty menopausal women were given kava for eight weeks, three times a day. Another group of menopausal women took a harmless pill called a placebo. After only one week, the women using kava showed improvement on psychological tests measuring anxiety. Their scores continued to improve through the rest of the eight weeks. Menopausal women using kava also reported improvement in their general mood and sense of well-being, and they had fewer hot flashes than women in the placebo group.

Q. Does kava have a healing effect on the bladder?

A. Most of the evidence that kava can heal the urinary tract from painful cystitis and infections is based on folklore. No major studies have been done to test its validity, but kava has been used for centuries for this purpose. Based on its historical use and relative safety, it's certainly worth giving kava a try for several weeks.

5.

Taking Kava

Like all plants, kava contains a distinctive complex of biologically active compounds. These compounds are why kava "works." This chapter will explain some appropriate ways to take kava, so you maximize the benefits of this Polynesian herb. It will also discuss the safety of kava use, including possible side effects and drug interactions.

Q. What's the best form to take kava in?

A. Kava supplements are available as capsules (powdered extract in a capsule that dissolves in the stomach) and tinctures (liquid extract in an alcohol base). Most of the clinical studies have used capsules, and there's no indication that tinctures work any better or worse. Kava has a numbing effect on the mouths of those who chew or drink it, and its taste has been described as soapy and bitter—not exactly a treat. If you want to avoid the taste or numbing effects of the tincture, use capsules. If you have trouble swallowing pills, drop the recommended dose of tincture into a few tablespoons of water or juice and drink it down.

Q. Should kava be taken on an empty stomach?

A. Although kava has a more intense effect if taken without food, taking it with meals won't decrease its absorption. Unlike some vitamins, it doesn't seem to require any special kinds of food to be absorbed. It's safe either way, so you can try it both with food and between meals to see which works best for you.

Q. Does kava have any side effects?

A. If kava is taken in high doses—as it sometimes is by South Pacific Islanders and Australians who drink kava beverages every day for long periods—a skin condition called *kava dermopathy* can develop. When 310 to 440 grams (that's 310,000 to 440,000 mgs, or almost a pound!) are consumed per week for a few months, the skin becomes scaly and flaky, especially on the palms of the hands, soles of the feet, forearms, back, and shins. Other consequences of kava abuse are liver damage, breathing problems, and unhealthy changes in blood cell counts. All of these effects are easily reversed by reducing kava consumption.

Q. Is there anyone who should not take kava?

A. Women who are pregnant or nursing and anyone using prescription antidepressants or antianxiety drugs should not use kava. Kava can worsen symptoms of Parkinson's disease, so if you have been diagnosed with this brain disorder, don't use kava. In addition, infants and children should not be given kava, since kava's effects on them are not known.

Q. Does kava interact with any prescription medications?

A. Yes, it does. You should not take kava if you are using prescription medications for depression or anxiety. If you have been using these medications, you should work with your physician to wean yourself off them completely before trying kava. The effects of these drugs will be increased to potentially dangerous levels if you use kava with them.

It's important that you not abruptly stop these medications, because your body grows accustomed to them and needs them to function after you've taken them for a while (like an addiction). All of these medications change the balance of potent neurotransmitters in your brain. Decreasing them gradually gives you time to adjust to the changes in body chemistry.

Q. Specifically, which drugs should I not take with kava?

A. Do not use kava if you are taking any of the following drugs:

Antianxiety Drugs

- BuSpar (buspirone)
- Trancopal (chlormezanone)
- Atarax, Vistaril (hydroxyzine)
- Ativan (lorazepam)
- Equanil, Miltown (meprobamate)
- Xanax (alprazolam)
- Lectopam (bromazepam)
- Librium, Libritabs (chlordiazepoxide)
- Klonopin (clonazepam)
- Tranxene (clorazepate)
- Valium, Vazepam (diazepam)
- Dalmane (flurazepam hydrochloride)
- Paxipam (halazepam)
- Loftran (ketazolam)
- Ativan (lorazepam)
- Versed (midazolam hydrochloride)
- Mogadon (nitrazepam)
- Serax (oxazepam)

- Centrax (prazepam)
- Doral (quazepam)
- Restoril (temazepam)
- Halcion (triazolam)

Antidepressant Drugs
- Prozac (fluoxetine hydrochloride)
- Effexor (venlafaxine hydrochloride)
- Elavil, Endep (amitriptyline hydrochloride)
- Asendin (amoxapine)
- Anafranil (clomipramine hydrochloride)
- Norpramin, Pertofrane (desipramine hydrochloride)
- Adapin, Sinequan (doxepin hydrochloride)
- Janimine, Tofranil (imipramine hydrochloride)
- Aventyl, Pamelor (nortriptyline hydrochloride)
- Vivactil (protriptyline hydrochloride)
- Surmontil (trimipramine maleate)

- Ludiomil (maprotiline hydrochloride)
- Wellbutrin (bupropion)
- Luvox (fluvoxamine maleate)
- Serzone (nefazodone hydrochloride)
- Paxil (paroxetine hydrochloride)
- Zoloft (sertraline hydrochloride)
- Desyrel (trazodone hydrochloride)
- Any monoamine oxidase inhibitor

Q. Can I drive while taking kava?

A. Gauge kava's effects on your alertness and reaction time before driving. The dose recommended to soothe anxiety and depression shouldn't affect your ability to drive safely. If you think kava makes you sleepy even at this low dose, use caution when you get behind the wheel and experiment with slightly lower dosages.

Q. Is it all right to drink alcohol while using kava?

A. There are no known dangerous interactions between alcohol and kava; however, if you do have an alcoholic drink while taking kava, be prepared to become intoxicated by much smaller amounts of alcohol. Do not try to drive or operate machinery of any kind if you've had alcohol on top of kava.

Q. Is kava addictive?

A. No, kava isn't physically addictive, and those who use it don't develop a tolerance that forces them to use more over time to reap the benefits. Theoretically, it's possible to become psychologically dependent on kava because of the way it makes you feel. It's so safe in low dosages, however, that you generally don't need to be concerned about long-term use.

Conclusion

If you look at the faces of people around you, you'll see many people who are unhappy, anxious, depressed, and tired. They are not enjoying the many possible joys of life. Some, in fact, are affected by mood-altering—and dangerous—prescription drugs.

Kava, a natural remedy used for centuries by the natives of Polynesia, is now available as a supplement for larger numbers of people. It can reduce anxiety and lift depression, and it can enhance health in many other ways.

Maybe kava sounds too good to be true. A completely natural substance with so many benefits and so few risks and so many uses certainly does give the impression that there ought to be a catch. So far, I haven't heard of one. If you follow the information and guidelines in *All About Kava*, and you do not use kava as an escape from problems that need your attention, you should only benefit from this amazing natural remedy.

Glossary

Anxiety. A feeling of worry, nervousness, or apprehension about the future. Although anxiety can be a normal response to situations, chronic and generalized anxiety can affect mental and physical health.

Depression. A feeling of sadness without any hope of feeling better. Although most people have periods of sadness or feeling "blue," they realize that these feelings will pass.

Hypersomnia. A condition characterized by excessive sleeping.

Insomnia. The inability to sleep, either because of difficulty falling asleep or staying asleep.

Kava. A plant indigenous to the Pacific islands whose leaves and roots are used to relieve such conditions as anxiety and depression.

Kavalactones. The naturally occurring active ingredients in kava.

Neurotransmitter. One of many chemicals found in the brain that enhances communication among brain cells and thinking and mood.

Serotonin. One of the brain's principal neurotransmitters, which promotes a relaxed, upbeat mood.

Serotonin reuptake inhibitors (SSRIs). A class of drugs that maintain high levels of serotonin in the brain.

Stress. An emotionally or mentally disruptive condition resulting from adverse external stimuli.

References

American Psychiatric Press, *Diagnosis and Treatment of Anxiety Disorders: A Physician's Handbook*, Washington DC, 1991.

Backhauss C, Krieglstein J, "Extract of kava *(Piper methysticum)* and its methysticin constituents protect brain tissue against ischemic damage in rodents," *European Journal of Pharmacology* 215 2–3 (May 14, 1992): 265–269.

Blazer D, George L, Hughes D, "The epidemiology of anxiety disorders: an age comparison," *Anxiety in the Elderly*, Springer/Verlag, New York (1991) 17–30.

Bone K, "Kava: a safe herbal treatment for anxiety," *Townsend Letter for Doctors* (June 1995): 84–87.

Bowers PJ, "Selections from current literature: psychiatric disorders in primary care," *Family Practice* 10(2) (1993): 231–237.

Davies LP, et al, "Kava pyrones and resin: studies on GABAA, GABAB and benzodiazepine binding sites in rodent brain," *Pharmacology and Toxicology* 71(2) (Aug 1992):120–126.

Elias M, "Depression, anxiety trigger high blood pressure," *USA Today* (September 4, 1997).

Foster S, "Kava kava," *Health Foods Business* (Dec. 1995) 40–41.

Frater AS, "Medical aspects of kava," Transactions and Proceedings of the Fiji Society 2 (1958): 31–39.

Gleitz J, et al, "Anticonvulsive action of (+/-)-kavain estimated from its properties on stimulated synaptosomes and Na+ channel receptor sites," *European Journal of Pharmacology* 315(1) (Nov 1996): 89–97.

Gleitz J, Beile A, Peters T, "(+/-)-Kavain inhibits veratridine-activated voltage-dependent Na(+) channels in synaptosomes prepared from rat cerebral cortex," *Neuropharmacology* 34(9) (Sep 1995): 1133–1138.

Hansel R, "Kava-kava in modern drug research: portrait of a medicinal plant," *Zeitschrift fur Phytotherapie* 17 (1996): 180–195. Translated by Clay A, Reichert R, for *Quarterly Review of Natural Medicine* (Winter 1996): 259–274.

Heinze HJ, et al, "Pharmacopsychological effects of oxazepam and kava-extract in a visual search paradigm assessed with event-related potentials," *Pharmacopsychiatry* 27(6) (Nov 1994): 224–230.

Holz HP, "The anxiolytic efficacy of the kava special extract WS 1490 using long-term therapy—a randomized, double blind study," presented at the 6th Phytotherapy Congress, *Quarterly Review of Natural Medicine* (fall 1996): 185–186.

Jamieson DD, Duffield PH, "The antinociceptive actions of kava components in mice," *Clinical and Experimental Pharmacology and Physiology* 17(7) (Jul 1990): 495–507.

Jamieson DD, et al, "Comparison of the central nervous system activity of the aqueous and lipid extract of kava *(Piper methysticum),*" *Archives Internationales de Pharmacodynamie et de Therapie* 301 (Sep 1989): 66–80.

Jussofie A, Schmiz A, Hiemke C, "Kavapyrone enriched extract from *Piper methysticum* as modulator of the GABA binding site in different regions of rat brain," *Psychopharmacology* (Berlin) 116(4) (Dec 1994): 469–474.

Kilham C, "Kava: a review," *Pure World Botanicals* (1996).

Kowachi I, et al, "Symptoms of anxiety and risk of coronary heart disease: the Normative Aging Study," *Circulation* 90 (Nov 1994): 2225–2229.

Lanza FL, "A review of gastric ulcer and gastroduodenal injury in normal volunteers receiving aspirin and other nonsteroidal anti-inflammatory drugs," *Scandinavian Journal of Gastroenterology* Suppl. (Norway) 42(163) (1989): 24–31.

Lehmann E, Kinsler E, Friedemann J, "Efficacy of special kava extract *(Piper methysticum)* in patients with states of anxiety, tension, and excitedness of non-mental origin—a double blind, placebo controlled study of four weeks treatment," *Phytomedicine* Vol. III (2) (1996): 113–119.

Lemonick MD, "The mood molecule," *Time* 50(13): (Sept. 29, 1997).

Magura EI, et al, "Kava extract ingredients, (+)-methysticin and (+/-)-kavain inhibit voltage-operated Na(+) channels in rat

CA1 hippocampal neurons," *Neuroscience* 81(2) (Nov 1997): 345–351.

Munte TF, et al, "Effects of oxazepam and an extract of kava roots *(Piper methysticum)* on event-related potentials in a word recognition task," *Neuropsychobiology* 27(1) (1993): 46–53.

Murray MT, N.D., *Natural Alternatives to Prozac,* William Morrow and Company, New York; 1996.

Russell PN, Bakker D, Singh NN, "The effects of kava on altering speed of access of information from long term memory," *Bulletin of the Psychonomic Society* 25(4) (1987): 236–237.

Seitz U, et al, "Relaxation of evoked contractile activity of isolated guinea-pig ileum by (+/-)-kavain," *Planta Medica* 63(4) (Aug 1997): 303–306.

Singh YN, "Effects of kava on neuromuscular transmission and muscle contractility," *Journal of Ethnopharmacology* 7(3) (May 1983): 267–276.

Singh YN, Blumenthal M, "Kava: an overview," *Herbalgram—the Journal of the American Botanical Council and the Herb Research Foundation* No. 39 (Spring 1997): 33–56.

Volz HP, Kieser M, "Kava-kava extract WS 1490 versus placebo in anxiety disorders—a randomized placebo-controlled 25-week outpatient trial," *Pharmacopsychiatry* 30(1) (Jan 1997): 1–5.

Walton J, et al, "Effects of kawain and dihydromethysticin on field potential changes in the hippocampus," *Progress in Neuropsychopharmacology Biology, and Psychiatry* 21(4) (May 1997): 697–706.

Weiss KJ, "Confronting anxiety and depression in primary care," *New Jersey Medicine* 91(3) (1994): 157–158.